Samantha Cardenas

Genetic Mapping for Bipolar Disorder and Schizophrenia

Samantha Cardenas

Genetic Mapping for Bipolar Disorder and Schizophrenia

Biomedical Protocol for Bipolar Disorder and Schizophrenia

LAP LAMBERT Academic Publishing

Impressum / Imprint
Bibliografische Information der Deutschen Nationalbibliothek: Die Deutsche
Nationalbibliothek verzeichnet diese Publikation in der Deutschen Nationalbibliografie;
detaillierte bibliografische Daten sind im Internet über http://dnb.d-nb.de abrufbar.
Alle in diesem Buch genannten Marken und Produktnamen unterliegen warenzeichen-,
marken- oder patentrechtlichem Schutz bzw. sind Warenzeichen oder eingetragene
Warenzeichen der jeweiligen Inhaber. Die Wiedergabe von Marken, Produktnamen,
Gebrauchsnamen, Handelsnamen, Warenbezeichnungen u.s.w. in diesem Werk berechtigt
auch ohne besondere Kennzeichnung nicht zu der Annahme, dass solche Namen im Sinne
der Warenzeichen- und Markenschutzgesetzgebung als frei zu betrachten wären und
daher von jedermann benutzt werden dürften.

Bibliographic information published by the Deutsche Nationalbibliothek: The Deutsche
Nationalbibliothek lists this publication in the Deutsche Nationalbibliografie; detailed
bibliographic data are available in the Internet at http://dnb.d-nb.de.
Any brand names and product names mentioned in this book are subject to trademark,
brand or patent protection and are trademarks or registered trademarks of their respective
holders. The use of brand names, product names, common names, trade names, product
descriptions etc. even without a particular marking in this works is in no way to be
construed to mean that such names may be regarded as unrestricted in respect of
trademark and brand protection legislation and could thus be used by anyone.

Coverbild / Cover image: www.ingimage.com

Verlag / Publisher:
LAP LAMBERT Academic Publishing
ist ein Imprint der / is a trademark of
OmniScriptum GmbH & Co. KG
Heinrich-Böcking-Str. 6-8, 66121 Saarbrücken, Deutschland / Germany
Email: info@lap-publishing.com

Herstellung: siehe letzte Seite /
Printed at: see last page
ISBN: 978-3-659-58121-2

Zugl. / Approved by: Richmond, Eastern Kentucky University, 2014

Acknowledgements:

I would like to give special thanks to my family for supporting me through the process of writing this dissertation and throughout my continued education. I give an extended thanks to my mom, Kendall, who was encouraging and being continually supportive. I would also like to thank Dr. Jerome May for taking on the role as mentor for this project.

Table of Contents:

Introduction

Bipolar Disorder (BPD) and Schizophrenia are believed to be among the top ten disabilities worldwide and are thought to affect about 2% of the population (1% each). Some of the most creative literary minds of the 20th century were known to be afflicted with Bipolar Disorder. But while this possible link to creativity is intriguing, the focus of this research paper will be to trace the benefits of tried-and-true conventional treatments, as well as the promise of novel genetic techniques, as scientists attempt to identify the genes responsible for BPD. Currently, the diagnosis of psychiatric disorders such as Schizophrenia and Bipolar Disorder is based predominantly on their clinical indices. Several different methods have been investigated as a way to diagnose Bipolar Disorder or Schizophrenia with a single genetic test; but there has been inconsistent success, so the search continues. While psychiatrists can establish the presence of an illness with relative ease, the process to differentiate between possible diagnoses based on symptoms of the patient can be very complicated, especially in the early phases of Schizophrenia and Bipolar Disorder[1]. During the process to find the solution, many new insights have surfaced. It has been discovered through research that similar genes may be responsible for both BPD and Schizophrenia[2]. In the meantime, patients with Bipolar Disorder (and sometimes Schizophrenia) are prescribed lithium. Although lithium's mechanics for alleviating BPD symptoms is not completely known yet, it still

---

[1] Schnack et al., 2014, p. 136
[2] Turecki et al., 2001, p. 136

remains the best treatment today. The discovery of genes causing BPD and other psychiatric disorders will almost certainly lead to major advances in understanding their pathogenesis, which will then allow researchers to find better and more effective treatments[3]. Most scientists and researchers in this field are interested in identifying the biological characteristics of mental illness and have high hopes to soon find the underlying susceptible genes for psychiatric disorders. The most important step we can take is to understand these disorders, because the disease will be seen at a level that allows for personalized medical treatment. This will potentially lead to improved treatment options[4].

---

[3] Lachman, 2000, p. 140
[4] Lachman, 2000, p. 136

What is Bipolar Disorder? A Primer

Bipolar Disorder, also known as manic depressive illness, affects approximately 1% of the population. BPD is also in the top ten leading causes of disability in the world[5]. By its very nature, BPD manifests in numerous parts of the brain, and these features can show up across a number of domains. The domains most affected include mood, energy, cognition, and personality[6]. Mounting evidence supports the disagreement that, although BPD is clearly not a classic neurodegenerative disorder, it is complemented by local brain volumetric diminuitions and cellular atrophy or cellular death[7]. A typical cycle for BPD is a manic cycle that can last for weeks and is then followed by weeks to even months of depression, with alternating cycles of mania and depression. The illness as described in the preceding sentence is referred to as Bipolar Disorder Type I; but with the cycling of hypomania (as opposed to a manic cycle) and depression, the illness is referred to as Bipolar Disorder Type II -- and in these patients it appears that the two moods cycle together[8]. Clinically, BPD II is found to be more common in females and has a slightly later onset than males most often diagnosed with BPD I[9]. Depending on the individual, cycles can occur approximately annually or semi-annually, once every few months (rapid cycling), weekly to monthly (ultra-rapid cycling);

---

[5] Lachman, 2000, p. 135
[6] Malhi et al., 2012, p. 67
[7] Shaltiel, Chen, & Manji, 2007, p. 25
[8] Lachman, 2000, p. 135
[9] Malhi et al., 2012, p. 70

or, very rarely, even every 24-48 hours (ultra-ultra-rapid cycling). The cycles of mania and depression have very severe symptoms and can become life threatening.

An individual going through a manic phase is characterized to exhibit an increase in energy and the feeling of grandiosity, power, and omnipotence. A manic phase can last anywhere from a few days to several weeks and can be hard to distinguish from Schizophrenia if auditory hallucinations and paranoia incapacitate the individual. Other symptoms of a manic cycle include excessive talking, egocentrism, and impulsivity. In manic stages, individuals are more likely to put themselves in dangerous or criminal situations. These individuals also propose often improbable plans in excessive numbers; thus leading them to feeling overwhelmed by the sheer volume of plans and ideas, and thereby decreasing productivity. Certain careers – politics, Science, business, and the arts -- benefit from this manic stage[10]. But because the clinical features of ADHD overlap with the symptoms of mania and hypomania, accurate diagnosis of mania can be difficult to achieve if the patient has ADHD[11]. Without a manic cycle, the diagnosis of Bipolar Disorder is no longer considered an accurate diagnosis. As described above, these are the characteristics of mania, but depression has a completely separate set of criteria to be met[12].

---

[10] Lachman, 2000, p. 140
[11] Lachman, 2000, p. 139
[12] Lachman, 2000, p. 135

The main characteristics of the depression cycle for someone suffering from Bipolar Disorder include intense sadness, hopelessness, anhedonia, deceased energy, and a lack of motivation or even caring about their own well-being. The individual in this stage can also have trouble focusing on school or work because of the loss of interest and lack of energy to care or give effort. Furthermore, there is an increase in feelings of worthlessness, worry, and anxiety; thoughts of suicide are probable. One major way to distinguish between the depressive phase of Bipolar Disorder and that of major depressive disorder, (a.k.a., unipolar depression) are sleeping and eating habits. BPD patients sleep more and eat more carbs, whereas those suffering from major depressive disorder (no manic cycles) have insomnia and eat less. Although the symptoms are easily identifiable; the cause has yet to be pinpointed.

Variations in the phenomenology of mood can sometimes be attributed to additional complications that can frequently mask the correct diagnosis or make a diagnosis and treatment harder. These complications can include conditions such as substance abuse and anxiety disorders; and these comorbidities have been linked with poorer recovery, reduced quality of life, and a heightened risk of suicide[13]. Anxiety disorders (63%) were the most commonly reported comorbidities, followed by behavioral (45%), and substance abuse disorders (37%). Many times, the behavioral problems are due to the BPD or other complicating factors. Comorbid anxiety disorders present a treatment challenge for clinicians, because the anxiety hampers treatment

---

[13] Malhi et al., 2012, p. 67

response. BPD comorbid with anxiety really accentuates results of a poorer quality of life, elevated suicide risk, and increased health service utilization[14]. When diagnosing someone with Bipolar Disorder, scientists and doctors look at IQ and school performance, because high numbers in both can be predictive for Bipolar Disorder[15].

The Causes of Bipolar Disorder

Twin, family, and adoption studies validate the notion that genetic factors play a large role in Bipolar Disorder susceptibility. In monozygotic twin studies, the concordance rate of BPD is: if one twin has BPD, there is a 60-80% chance that the other twin will also have BPD[16]. Bipolar Disorder has been found to be a polygenic disease - a disease linked with environmental factors that contribute to the onset of one's illness and also linked with multiple genes in amalgamation with lifestyles[17]. Since the concordance between monozygotic twins in not 100%, it is a signal that the phenotype being studied, in this case, BPD is not based solely on genetics[18]. Bipolar Disorder has always been recognized to be linked with genetics; and because of this known fact, numerous studies have been carried out in attempts to map the susceptibility loci. These studies have generated various possible regions with potential loci, although none of the studies have identified the exact susceptibility genes. Recent meta-analysis of genome scans presented evidence for the most probable location, but

---

[14] Malhi et al., 2012, p. 74
[15] Demjaha, MacCabe, & Murray, 2011, p. 210
[16] Lachman, 2000, p. 139
[17] Offord, 2012, p. 134
[18] Offord, 2012, p. 133

all established genes established were unpersuasive. It is difficult to locate the specific genes causing BPD because of the illness' inherent intricacy within one's genetic mechanism. Bipolar Disorder is classified to express both heterogeneous and varying phenotype expression. In summary, no single study so far has provided strong enough conclusions to be considered the 'best option' in the advancement in identifying vulnerable genes for the traits. It has been agreed upon that a comprehensive approach using a number of different methodologies will be needed in the hopes of diagnosing and lessening the damages of BPD[19].

Treatment of Bipolar Disorder

Maintaining Bipolar Disorder can sometimes be achieved through medication. These medications can include mood stabilizers and anti-manic drugs such as lithium, valproate, and carbamazepine. The administration of antipsychotics and/or antidepressants in combination with the mood stabilizers may also be used as a treatment strategy. Although some patients' symptoms can be reduced with medication, about one-third of patients do not see the anticipated progress and are still unable to lead normal lives. However, some patients voluntarily stop taking the prescribed medications, either because of the side effects from the medicine they take such as uncontrollable twitches; or because, similar to someone with an infection who quits

---

[19] Alda, Grof, Rouleau, Turecki, & Young, 2005, p. 1039

taking their prescription once they begin to feel better, BPD patients sometimes stop taking their medications as soon as they return to their baseline[20].

Lithium

Mood stabilizers are chemical compounds used to stabilize euthymic patients by helping to prevent those patients from experiencing a manic and/or depressed state[21]. Mood stabilizers are used for BPD and Schizophrenia. Mood stabilizers are often prescribed along with antidepressants to prevent mood swings. Anticonvulsants such as lamotrigine, carbamazephone, valproic acid, or lithium are frequently used as mood stabilizers because they treat both problems (not only seizures but also mood swings)[22]. Bolstered by 50 years of successful outcomes, lithium remains as the number one choice for treating mood disorders today[23]. Lithium is very effective and is the most widely used anticonvulsant and has been used as a mood stabilizer since the late 1940s. This is because lithium has been continually studied and consistently shows effectiveness as a treatment for some psychiatric disorders. But due to lithium's therapeutic index – the ratio between the dosage of a drug that causes a lethal effect and the dosage that causes therapeutic effects -- close to one, lithium can be very difficult to monitor. Lithium's effectiveness is of major consequence to whether the BPD is more genetically-caused or more environmentally-caused. Research has shown that

---

[20] Lachman, 2000, p. 138
[21] Offord, 2012, p. 134
[22] Offord, 2012, p. 135
[23] Lachman, 2000, p. 135

lithium is more effective in people whose psychosomatic ailment follows a periodic course, has low rates of comorbidity, lacks rapid cycling, and is accompanied by the patient has a family history of Bipolar Disorder. A study on lithium-responders verses non-responders has also provided scientists with evidence that treatment must differ for the different neuroendocrine systems between responders and non-responders. Signal transduction pathways play an important role for the pathophysiology of Bipolar Disorder. These signal transduction pathways are also used as the assumed target pathway for lithium to start working against the chemicals in the brain triggering the disorder[24]. Over the past twenty-plus years, multiple studies have supported that lithium-responsive patients are more likely to have blood-relatives that have also been diagnosed with BPD than those who saw no change in mood when taking the lithium. Due to the correlation for the lithium-responsive patients, it has been concluded that there is "a major gene effect consistent with an autosomal recessive mode of inheritance." With these facts in mind, this observation could eventually lead scientists to the answer about the distinction between a Bipolar Disorder phenotype with less genetic heterogeneity and a different genotype that is affected more and by stronger genetics[25,26]Scientists are more reliant on those with Bipolar Disorder that have a positive response to lithium, because they believe the answer is very likely to come from studying their genetic maps. This is because lithium responders are a collection of

---

[24] Offord, 2012, p. 134
[25] Lachman, 2000, p. 135
[26] Alda, Grof, Rouleau, Turecki, & Young, 2005, p. 1039

people with distinctive medical features that correspond to the illness' "core phenotype." Hence, their family histories have allowed scientists to work within a smaller phenotype spectrum. Concluding that one's response to lithium prophylaxis is hereditary, it was determined that lithium responders are a more homogenous subclass of Bipolar Disorder. The individuals in this subclass are characterized by having a "strong genetic loading" and "a hereditary conduction harmonizing with a major-gene effect."

Lithium responders also differ from responders to other mood stabilizers such as valproate, carbamazepine, lamotrigine, or olanzapine; lithium-response appears to be specific. The first sign of this distinction was seen when anti-manic responses to valproate and lithium were noted to be different between patient groups in studies. Secondly, the Germany MAP study, a major trial of prophylactic treatment of Bipolar Disorder, was able to show that, unlike lithium responders, patients who advanced from long-term treatment with carbamazepine had unusual clinical features when compared to the lithium responders. Lastly, when comparing lithium responders to lamotrigine users, there were differences identified concerning comorbid conditions and their family history. Patients that showed a positive response to lamotrigine had higher rates of comorbid conditions, especially anxiety, as well as a family history of anxiety disorders. In contrast to lamotrigine responders, lithium responders are characterized to have had higher rates of BPD in relatives[27]. In conclusion, based on genetic heritability, children of lithium responders are at a higher risk for developing mainly Bipolar Disorder or

[27] Alda, Grof, Rouleau, Turecki, & Young, 2005, p. 1039-1040

frequent depression, while offspring of non-responders have displayed a broader and less specific array of psychiatric conditions[28].

What is Schizophrenia?

Schizophrenia is usually described as a disease of young people due to its onset primarily in people in their late teens or early twenties. Its victims often feel as if they have lost who they are; with reports of 'losing' their identity, autonomy, and mental capacity. Ten percent of people with Schizophrenia die by suicide. Not only do the diagnosed fall victim to this neurodegenerative disease, but also the loved ones around the patient. Parents of children with Schizophrenia have described that this disorder seems to take the children they knew and turns them into completely different people[29]. Schizophrenia is an incapacitating mental illness characterized by extensive and visible brain abnormalities[30].

Just as there are different times people can be diagnosed with cancer Schizophrenia can too arise at different times in life. Adult-onset Schizophrenia usually occurs during late adolescence and young adulthood. Adult-onset Schizophrenia is also much more common than childhood-onset Schizophrenia[31]. Childhood-onset Schizophrenia is an uncommon and is a more severe form of the disorder characterized by the onset of psychosis before the age of 13; the disorder neurobiologically continues

---

[28] Alda, Grof, Rouleau, Turecki, & Young, 2005, p. 1041
[29] Andreasen, 2005, p. 2
[30] Moran, Pol, & Gogtay, 2013, p. 3215
[31] Moran, Pol, & Gogtay, 2013, p. 3219

to degrade as the patient ages into adulthood[32]. Childhood-onset Schizophrenia causes a profound and progressive cortical grey matter damage that continually spreads throughout one's parieto-frontal and parieto-temporal lobes during adolescence. As the disorder advances, the degeneration mainly focuses on the prefrontal and temporal cortices as children age into young adults[33]. Neurobiological research suggests that child-onset Schizophrenia and adult-onset Schizophrenia share common originating factors, although child-onset Schizophrenia is usually considered to be a more severe form[34]. It is an accepted fact that people with established Schizophrenia show brain structure irregularities[35]. When looking at a patient with Schizophrenia overall, a few patterns stick out -- such as, exhibiting lower IQ, suffering from memory and language impairment, and having poor executive functioning when compared to healthy individuals[36].

Diagnosis of Schizophrenia

The diagnosis of Schizophrenia is the same for both child-onset and adult-onset Schizophrenia. The criterion to be diagnosed with Schizophrenia is outlined in the DSM-V. The DSM-V requires that two or more characteristic symptoms -- i.e., hallucinations, delusions, disorganized speech, disorganized or catatonic behavior, and/or negative symptoms -- must be present for at least one month. During the active phase,

---

[32] Moran, Pol, & Gogtay, 2013, p. 3216
[33] Moran, Pol, & Gogtay, 2013, p. 3217
[34] McClellan & Weery, 2013, p. 978
[35] Demjaha, MacCabe, & Murray, 2011, p. 209
[36] Demjaha, MacCabe, & Murray, 2011, p. 211

hallucinations, delusions, or disorganized speech must be present. Evidence of the

disorder must be present for at least six months and must be associated with a

significant decline in social or occupational functioning. In children and adolescents,

decline in function may include the failure to achieve age-appropriate levels of

interpersonal or academic development[37]. There are trademark phases present in

Schizophrenia that are very important to recognize when one is making diagnostic and

therapeutic decisions. Most patients exhibit some degree of deterioration in day-to-day

functioning. This deterioration is present before the onset of the psychotic symptoms.

The deterioration phase usually includes social withdrawal or isolation, eccentric or

bizarre preoccupations, behavior unusual for the patient, new academic failure,

deteriorating self-care skills, and/or dysphoria[38].

The critical phase for the beginning of Schizophrenia is marked by prominent

positive symptoms and substantial declines in personal functions. The next step is the

recuperative/recovery phase. This occurs after the acute phase and has signs of

remission of the acute psychosis. Usually, in the recovery phase, there is a several

month period when the patient still does not continue to function at their base line; both

negative and positive symptoms can be present[39]. The final stage is the residual phase.

---

[37] McClellan & Weery, 2013, p. 977
[38] McClellan & Weery, 2013, p. 979
[39] McClellan & Weery, 2013, p. 979

At this time in the cycle there are periods recognized between acute phases when the affected person does not experience significant positive symptoms[40].

Symptom of and Brain Abnormalities Associated with Schizophrenia

The signs and symptoms of Schizophrenia are diverse and may include: perceptive disorders (hallucinations), inferential thinking (delusions), goal directed behavior (avolition), and emotional expression (affective blunting)[41]. Disorganized behavior is also very common, and a patient may express disorganized speech, bizarre behavior, and a poor attention span[42]. None of the above symptoms alone are enough to diagnose someone with Schizophrenia; also, none of the above symptoms pertain only to Schizophrenia[43]. Positive and negative symptoms are associated during the different phases of Schizophrenia can be distinguished. Positive symptoms are side effects such as hallucinations, delusion, and thought disorders. Negative symptoms are characterized by being deficits like low energy, lack of emotion, and slowed speech or thought[44]. Years of research and evidence conclude that Schizophrenia is a progressive brain disease[45]. The molecular mechanism of the brain abnormalities found in Schizophrenia remains unclear at this point in time. By trying to understand

---

[40] McClellan & Weery, 2013, p. 980
[41] Andreasen, 2005, p. 3
[42] McClellan & Weery, 2013, p. 981
[43] Andreasen, 2005, p. 9
[44] McClellan & Weery, 2013, p. 982
[45] Van Haren, Cahn, Hulshoff Pol, & Kahn, 2014, p. 11

Schizophrenia's molecular mechanism, scientists could conclude whether the brain abnormalities are due to the illness' state itself, or if it is more tied to genetic risk[46].

Schizophrenia is known to have extensive gray matter reductions in the affected brains; abnormal brain structure is one of the most vigorous biological features of Schizophrenia[47]. Data suggest that the largest decline in volume of the grey matter takes place during the first year of onset of the illness[48]. The most consistent findings associate the brain abnormalities with the reduction of grey matter in the prefrontal and temporal cortices and the medial temporal lobe structure; but at the same time, there is an increase in the lateral ventricular volume. In patients with typical adult-onset Schizophrenia, research has shown that progressive brain degeneration of patients' brain tissue is double-fold compared to healthy control subjects. This brain reduction is even more severe in cases of childhood-onset Schizophrenia[49]. One's hippocampus amygdala is also seen to be reduced by the neurodegenerative disease[50,51]. Grey matter is not the only place differences are found between healthy patients and Schizophrenia patients. Neuroimaging studies have revealed that white matter, too, is affected; and the longer one has had the illness, the more pronounced their deficits in

---

[46] Moran, Pol, & Gogtay, 2013, p. 3216
[47] Demjaha, MacCabe, & Murray, 2011, p. 211
[48] Van Haren, Cahn, Hulshoff Pol, & Kahn, 2014, p. 11
[49] Moran, Pol, & Gogtay, 2013, p. 3215
[50] Demjaha, MacCabe, & Murray, 2011, p. 210
[51] Haren et al., 2012, p. 915

their white matter[52]. But in the end, post-mortem studies have revealed that the largest

impact Schizophrenia has on the brain lies in the dorsolateral prefrontal cortex

comparative to other cortical regions[53]. Siblings of those with Schizophrenia have been

found to have brain scans that reveal that their brain image often has characteristics of

both Schizophrenic brains and normal, healthy brains. The major difference is the lack

of brain degeneration compared to those with full-blown Schizophrenia[54].

Causes of Schizophrenia

A Schizophrenia patient's multifactorial predispositions that impair

neurodevelopment are genetics and environmental factors[55,56]. Schizophrenia has a

heritability rate of 85%[57]. The timing of the interaction between environmental factors

and biologic risk factors facilitates the time of onset, the course of the disease, and

severity in people with Schizophrenia[58]. Some studies have found a link between social

withdrawal at birth and Schizophrenia. Obstetric events are frequently reported to

facilitate on increased risk for Schizophrenia[59].

It is now well established that obstetric impediments lead to a person more likely

to develop Schizophrenia. Three kinds of obstetric complications that lead to the

---

[52] Moran, Pol, & Gogtay, 2013, p. 3215
[53] Van Haren, Cahn, Hulshoff Pol,&d Kahn, 2014, p. 9
[54] Moran, Pol, & Gogtay, 2013, p. 3222
[55] Demjaha, MacCabe, & Murray, 2011, p. 209
[56] McClellan & Weery, 2013, p. 978
[57] Haren et al., 2012, p. 915
[58] McClellan &Weery, 2013, p. 979
[59] Demjaha, MacCabe, & Murray, 2011, p. 210

increased risks are babies that have fetal growth retardation, fetal perinatal hypoxia, and prenatal complications. Lower birth weight, small head circumference, and an overall smaller size for the gestational age has been associated with individuals who later develop Schizophrenia more so than babies born of an average birth weight. Together, these factors point toward fetal growth retardation being an underlying factor for Schizophrenia. It has been claimed that fetal growth retardation is facilitated by genetic effect, "as mothers with Schizophrenia also have higher rates of low birth weight among offspring."[60] Stillbirths and fetal or neonatal deaths occur considerably more often among schizophrenic mothers giving birth. Fetal growth retardation has been recognized as one of the earliest indicators of the neurodevelopmental course of Schizophrenia. Areas of development that are monitored include changes in motor, language, and cognitive aptitude from infancy into childhood. It is well known that individuals with Schizophrenia are more likely to experience numerous obstetric complications involving a lack of or a less-than-needed amount of oxygen reaching the individual's tissues; babies with oxygen deficiency are more apt to develop Schizophrenia. The hypothesis is that when oxygen deprivation occurs, this leads to reductions in gray-matter volume in one's cortical and subcortical regions, especially the hippocampus. The effects of fetal hypoxia are two-to-three-times greater for people born small for their gestational age. This further complicates the effects of hypoxia, as

---

[60] Clarke, Harley, & Cannon, 2005, p. 2

well as shows a connection between genotype and obstetric events[61]. It is possible that there is a stronger correlation than thought between obstetric events and Schizophrenia, since not all obstetric complications are recorded on birth records.

Prenatal complications are also being proved to increase the risk of one's developing Schizophrenia. Such complications can include prenatal stress, intrauterine malnutrition, and prenatal infection -- although the list of prenatal risk factors associated with Schizophrenia is quickly becoming overbearing to doctors and mothers. The reason prenatal complications are so detrimental to the fetus is because they increase risks for genetic mutations. Obstetric difficulties undeniably play a role in the etiology of Schizophrenia, but the how and the amount of influence these difficulties present is still somewhat unknown[62]. When it comes to the cause of Schizophrenia, no single reason has been identified due to the lack of evidence of this complicated neurodevelopment disease[63]. All in all, Schizophrenia is a heterogeneous disorder caused by many etiologies[64].

Treatment of Schizophrenia

Antipsychotic drugs are the course of treatment for people suffering from Schizophrenia. These medicines have been around since the mid-1950s. Antipsychotics do not cure the disorder, but rather lessen the symptoms suffered by the

---

[61] Clarke, Harley, & Cannon, 2005, p. 3
[62] Clarke, Harley, & Cannon, 2005, p. 5
[63] McClellan & Weery, 2013, p. 977
[64] McClellan & Weery, 2013, p. 978

afflicted. Patients also seem to function better, have a higher quality of life, and have an overall better outlook about life. The brand and dose of medicine varies from person-to-person in order to control the disorder as effectively as possible. The first antipsychotic drug was accidentally discovered and happened to be useful in the treatment of Schizophrenia. First came Thorazine, and then products such as Haldol, Prolixin, Navane, Loxapine, Stelazine, Trilafon, and Mellaril. These drugs are known as "neuroleptics - take the neuron" because, although they effectively treat positive symptoms, at the same time these medicines can cause "cognitive dulling and involuntary movements, among other side effects." Although the positive symptoms are addressed with antipsychotics, negative symptoms are not affected by the medicine. Now, as the 21st century commences, gene-mapping research is the best starting point for individualized drug development because of the high heritability rates of Schizophrenia and because of the up-and-coming genetic mapping techniques being applied[65].

Comparing Bipolar Disorder and Schizophrenia

Both Schizophrenia and Bipolar Disorder symptoms are typically first seen between adolescence and mid-twenties. Also, males develop the illnesses earlier than females by around 2.5 years. The best differentiating environmental factor relating to Schizophrenia and BPD is that Schizophrenia is more likely to be seen in people with an

---

[65] Roofeh, Tumuluru, Shilpakar, & Nimgaonkar, 2013, p. 16

urban upbringing[66]. Child abuse is highly customary in both, with a strong association between auditory hallucinations and childhood traumas in both disorders[67]. People with Schizophrenia seem to have fewer offspring than both BPD and healthy people, and BPD people have fewer babies than healthy individuals. Both disorders also have a high correlation between those affected and those who smoke tobacco and use illegal drugs, with a much stronger correlation between Schizophrenia and drugs. It is uncertain whether the drugs contribute to the onset of Schizophrenia or if affected persons use drugs to deal with the symptoms, or if it is a combination of both[68].

Returning to the idea that BPD and Schizophrenia are closely related by similar genetic mutations, researchers have found a number of genes encoding SNARE proteins and vesicle membrane proteins common to both disorders[69]. Since the 1950's when antipsychotics were first used for Schizophrenia, antipsychotic drugs have now been found to also be useful in treating BPD symptoms. Again, this is not surprising taking into account the dopamine dysregulation in both disorders[70]. At the same time, mood stabilizers used for BPD are also now being used to treat negative symptoms in schizophrenia that are not addressed by antipsychotics[71]. Copy number variants (CNV), when the number of copies of a particular gene fluctuates from person-to-person, have been established to be found in excess in people with Schizophrenia, thus linking

---

[66] Demjaha, MacCabe, & Murray, 2011, p. 212
[67] Demjaha, MacCabe, & Murray, 2011, p. 210
[68] Demjaha, MacCabe, & Murray, 2011, p. 212
[69] Lachman, 2000, p. 140
[70] Demjaha, MacCabe, & Murray, 2011, p. 209
[71] Demjaha, MacCabe, & Murray, 2011, p. 212

Schizophrenia with autism, ADHD, and epilepsy. But in contrast, CNVs are not found in high number in those afflicted with Bipolar Disorder[72].

Human Genome Project

The Human Genome Project (HGP) was the basis for dramatic genetic advancement. Without the HGP, the idea of personalized medicine would not have ever been plausible. The sequence developed by the Human Genome Project positively augmented biomedical research[73]. Using HGP's broad framework, it is possible for scientists to assemble fragmented information into full works of knowledge, useful for filling in the blanks about biological structures and functions[74].

Arguably, the greatest impact of genomics is the ability to examine biological occurrences in a complete, neutral, and hypothesis-free method. For medicine, genomics has allowed for the first efficient approaches to discovering underlying genes and cellular pathways for diseases[75]. Early experiments created epigenomic maps that showed the locations of specific DNA modifications, chromatin modifications, and protein-binding actions occurring across the human genome, but as technology has improved, research is focusing on sequencing different affected human samples in order to study mutations[76]. The future of genomics includes medical personnel having the ability to characterize patients' germline genomes. The purpose of this is to identify

---

[72] Demjaha, MacCabe, & Murray, 2011, p. 210
[73] Lander, 2011, p.1
[74] Lander, 2011, p. 2
[75] Lander, 2011, p. 3
[76] Lander, 2011, p. 5

strong, projecting transformations in order to allow for presymptomatic counselling, search for origins of diseases, and to detect heterozygous carriers. This research is currently focused on cancer genomes (by looking at somatic mutations), immune repertoires (by looking at B-cell and T-cell receptors), and microbiomes (by connecting microbial communities with diseases).

Research applications will contain characterizing genomes, epigenomes, and transcriptomes of humans as well other species[77]. During meiosis, epigenetic information is not completely erased, thereby allowing it to be transmitted through multiple replications. Epigenetic markers determine the timing and location of the gene expression. Due to epigenetic markers not being fully erased, they can be inherited and thus impact an individual's phenotype, disease vulnerability, and drug response. However, these DNA modifications that occur throughout a lifetime are not detectable by genotyping and therefore pose challenging limitations on personalized medicine based on DNA sequence variations[78].

DNA sequencing is being used more and more frequently in clinical settings in order to help doctors prescribe a treatment plan for unclear or undiagnosed diseases. Genomics is also being frequently used in cytogenetics labs. These experiments are using DNA microarrays' high levels of sensitivity to detect clinically important chromosomal imbalances. This will help bring about advances in diagnosing and

---

[77] Lander, 2011, p. 5
[78] Lander, 2011, p. 19

evaluating children with idiopathic developmental delays, major intellectual debilitations,

autism, and birth defects[79]. By comparing an afflicted person's genome sequence to

their parents' genomes, it will become possible to spot new, lethal mutations on the

rise. Because genome sequencing has becoming affordable, this process may become

very customary for parents trying to conceive a baby. Pediatricians will also use

genome sequencing to be able to better explain idiopathic disorders in children[80].

Compared to other diseases and disorders, the information related to genetic mapping

for psychiatric diseases is far less advanced and has a limited scope. Based on the

progress made towards personalized medicine, it has been determined that larger

genetic studies are needed in order to understand underlying cellular pathways for

psychological diseases[81]. The crucial objective for genetic sequencing is for the process

to become increasingly simple and cheap so that it can be habitually used as an all-

inclusive apparatus throughout biomedicine[82].

Potential Genetic Mapping Applications

Personalized medicine centers on the idea that an individual's unique genes can

play a substantial role in tailoring therapies. Characteristics studied include genetic

alterations, epigenetic modification, clinical symptomatology, observable biomarker

---

[79] Lander, 2011, p. 14
[80] Lander, 2011, p. 15
[81] Lander, 2011, p. 19
[82] Lander, 2011, p. 5

changes, and environmental factors[83]. Biomarkers are used as a way for researchers to study traits that reveal both genomic and environmental effects, and therefore represent the serious physiological state of an individual[84]. It is anticipated that neuroimaging will provide a valuable contribution toward biomarkers guiding personalized medical treatment[85]. Doctors will use genealogy in order to foresee disease susceptibility, create accurate diagnoses, and offer patients efficient and advantageous options that please both the doctor and patient[86]. A person's genes make them either more or less likely to respond to different treatment or medications. Genes can also help predict certain side effects a patient may experience from the medications[87]. There have been many successful cases using personalized medicine in oncology, but no reported success cases for psychology yet[88]. The future of genetic studies involves finding the link between susceptibility to a disorder and the response to a specific treatment, because the cornerstone of this research is the use of genomic biomarkers to exploit advances in prevention and treatment[89,90].

The conjugation of genomics with a drug treatment regimen is known as pharmacogentics. Pharmacogenomic involvement has been used to evade adverse

---

[83] Ozomaro, Wahlestedt, & Nemeroff, 2013, p.1
[84] Holsboer, 2008, p. 641
[85] Holsboer, 2008, p. 642
[86] Ozomaro, Wahlestedt, & Nemeroff, 2013, p.1
[87] Aldo et al., 2001, p. 570
[88] Ozomaro, Wahlestedt, & Nemeroff, 2013, p.1
[89] Alda, Grof, Rouleau, Turecki, & Young, 2005, p. 1042
[90] Bombard, Abelson, Simeonov, & Gauvin, 2013, p. 1197

effects and make sure that correct drug doses are prescribed[91]. The pharmacogenetics of depression, the study of different hereditary variants that determine which antidepressant drug should be used, has two goals. The first goal is to use genetic testing in order to help psychiatrists make medical decisions so that the quantifiable benefits of antidepressant treatment is maximized, while at the same time lessening adverse side-effects. Negative side effects include weight gain, suicidal thoughts, and erectile dysfunction. The second goal of pharmacogenetics for depression is to discover and examine candidate genes recognized to have a role in the pathophysiology of depression[92].

Advancement in biomarker development is due to neuroimaging and neuroendocrine profiles, proteomics, and metabolomics. Together, these strategies will result in tailored antidepressant treatments based on genetic and pathophysiological backgrounds. It is already recognized that depressed patients carrying the L allele experience better improvement when taking selective serotonin re-uptake inhibitors (SSRI) than patients carrying the S alleles. L-allele patients are also known to have better responses to treatment and higher remission rates. All antidepressants are broken down in the liver by cytochrome (P450) CYP isoenzymes. Activity of these isoenzymes depends on genetic variation and environmental factors like lifestyle and age. The difference in the CYP gene activity has led to evolving research for

---

[91] Bombard, Abelson, Simeonov, & Gauvin, 2013, p. 1197
[92] Holsboer, 2008, p. 637

personalizing the proper drug dose for a patient[93]. Identifying poor or ultrafast

metabolizers has become a vital tool in tailored drug treatment; it has shown promise

through use in oncology. Mutations in genes can determine a patient's response to a

given drug whether positive or negative[94].

Another layer of complexity involves epigenetic mechanisms, which is the study

of heritable changes in gene activity not triggered by modifications in one's DNA

sequence; these changes can stimulate gene expression through chemical chromatin

modifications[95]. The inferences of epigenetic modifications for antidepressants showed

that chronic stress reduces the activity of the enzyme histone deacetylase 5 (HDAC5).

HDAC5 inhibits gene activity by eradicating acetyl groups from histones, but it was

found that the effect on HDAC5 is reversed by continuous administration of an

antidepressant medication[96].

Adverse, early life experiences can end in an imprinted chemical modification

that continues into adulthood. This is an accepted explanation for the reason why

childhood traumas increase one's risk for mood disorders in adulthood. This fact

prominently validated that epigenetic modifications are responsive to drug intervention

and therefore rescindable. With this is mind, and tissue samples available, gene-

environment interactions can now be studied at the molecular level. The study of

---

[93] Holsboer, 2008, p. 638
[94] Holsboer, 2008, p. 640
[95] Holsboer, 2008, p. 644
[96] Holsboer, 2008, p. 641

monozygotic twins treated for major depression do not automatically elicit the same clinical response pattern because of epigenetic variances in both DNA-methylation patterns and histone acetylation accumulation during a lifetime[97]. Knowing this, it is easier for doctors to explain to patients why a medicine or treatment previously used for their recurrent major depression may not work the next time around[98].

Placing patients into subgroups based on their genotypes and biomarkers will fuel the road towards discovery for new treatments. In order to create these subgroups, personalized action for depression will depend on large numbers of people being studied as they try out new, multi-tiered procedures that are able to be replicated, and by biologically valid experiments in order to see progress and gain more acceptance of this practice. Gene function changes depend on both variant nucleotide sequences and copy number variations (CNV). Resulting from the completion of the Human Genome Project, it became widely known that a person's genome goes through stages of loss and gain of genetic material. Due to this, the weight of each copy number variation's contribution to human maladies is still uncertain. In cases of oncology where genotyping has had success, it is recognized that some malignancies are linked to heightened copy numbers of particular genes, but it is difficult to detect the epigenetic changes that have been occurring since birth through genotyping[99].

---

[97] Holsboer, 2008, p. 643
[98] Holsboer, 2008, p. 644
[99] Holsboer, 2008, p. 642

The most relevant topic for clinical research is the discovery and confirmation of biomarkers. Specific biomarkers will, in the long run, allow for the match up of the specific biomarker and a specific medication; identifying an individual's pathology is vital for this process to work. The extrapolation that a certain antidepressant will work effectively for a patient will allow for better treatment and lower costs for treatment, because there will be a lack of trial-and-error involved with finding a medicine that works for someone[100]. There are a few different ways researchers are trying to find all of the genes, biomarkers, etc., mentioned earlier. One major way researchers are looking for these markers are through the development of neuroimaging techniques, such as magnetic resonance (MR) or positron emission tomography (PET). The use of these tools has been able to answer the basic questions about neuroanatomy, neuropathology, neurophysiology, and cognitive neuroscience, thus allowing the science to progress as far as it has. An important strategy being used today is to try to link the clinical symptoms of Schizophrenia with the causal brain mechanisms[101]. The first approach to this strategy is to assume that current symptoms area caused by a certain area of the brain. The backbone to this assumption argues that Schizophrenia can be compared to diseases such as neruosyphilis or multiple sclerosis where the disease is amplified by multiple lesions in different areas of the brain causing the clinical symptoms. The second approach to this strategy is based on distributed parallel processing which theorizes that the fundamental abnormality for psychological

---

[100] Holsboer, 2008, p. 642
[101] Andreasen, 2005, p. 9

problems are due to problems in connectivity or wiring in the brain, and that specific symptoms reflect specific abnormal wiring. Relevant to this scheme, a solitary well-placed misconnection or damage to a certain brain circuit can produce a range of inter-related symptoms because of the intricate feedback loops being the basis of brain circuitry[102]. Science has revealed a short in the brain circuit responsible for filtering information taken in by the brain which then forwards the information or stimulant to the heteromodal cortex. A problem in this circuit induces feelings of being overloaded with information. With the brain being overrun by information, the end results can be hallucinations, delusions, disorganized behavior, and the negative symptoms discussed earlier. Validity of this proposed model include support by experiments that have demonstrated inadequate processing of information in Schizophrenia along with results pointing towards abnormal findings in the thalamus and related circuits[103].

Even though genetics is the primary cause for BPD or Schizophrenia, epigenetics also play a valid role on the onset of these disorders. Therefore, when studying personalized medication, epigenetics cannot be ignored. Epigenetics are genetic changes caused by sources that do not alter one's nucleotide sequence. While researching personalized medicine, scientists are realizing the true importance that environmental factors play in the role of psychological disorders. Mechanisms for the disorder's causes have to be re-investigated and revised to incorporate the outside

---

[102] Andreasen, 2005, p. 11
[103] Andreasen, 2005, p. 12

factors[104]. Incorporating transcriptomics, proteomics, and all proteins into studies will increasingly improve the baseline of information currently available to make well-versed resolutions in personalized psychiatry. In conclusion, multidisciplinary approaches are being used more frequently and will ultimately be the answer to personalizing medicine. Such combinations may inquire whether genes, environmental interactions, and/or endopheonotypes will produce distinctive phenotypic expression of the illness, thereby increasing the database to draw diagnoses from[105].

Medication (pharmacologic therapy) is not the only part of personalized medicine being studied. Other practices that are trying to be introduced to personalized medicine are various psychotherapies such as light therapy, deep-brain stimulation, electroconvulsive therapy, and transcranial magnetic stimulation. But it is still believed that combining neuroimaging with genetics methods will have the largest impact on personalized therapy, because the images studied can create measurable biological phenotypes. Neuroimaging incorporates genetics, psychiatry, and neuroscience that relate genetic variation back to a specific protein function, brain abnormality, and/or psychopathology. The two major neuroimaging genetic approaches currently being addressed are the identification of imaging changes in a group of people known to have a genetically focused illness and/or the confirmation of effects caused by the specific genetic change. A predominantly beneficial facet of describing biological phenotypes

---

[104] Ozomaro, Wahlestedt, & Nemerodd, 2012, p. 6
[105] Ozomaro, Wahlestedt, & Nemerodd, 2012, p. 15

through imaging is to look at the disease away from self-reported and possibly

unreliable diagnostics. The aptitude to study in vivo change and change over time for

psychiatric diseases is vital because of the already identified neuronal plasticity. All in

all, neuroimaging genetics is showing promising results and may be the necessary

insight to differentiate between the understanding of the pathophysiology of well-known

psychiatric disorders by assessing the connection between imaging phenotypes and

genetic variation[106].

Although neuroimaging has produced reliable data, there are some limitations to

this type of study. As soon as mediation, like antidepressants, for psychological

disorders are used and work, the changes in one's brain can be seen, possibly

revealing a cause. With neuroimaging genetics being a new area of study, there is a

higher possibility for false positives and imprecise interpretations. There is also some

skepticism that neuroimaging results being studied are actually showing the functional

change being investigated. In rebuttal, this type of research emphasizes that MR

spectroscopy can provide chemical profiles able to tell the difference between brain

pathologies.

Another strategy being used for neuroimaging is the use of positron emission

tomography/single-photon emission computed tomopraphy. This is more intrusive than

previously mentioned methods, but has the advantage of permitting in vivo monitoring

---

[106] Ozomaro, Wahlestedt, & Nemerodd, 2012, p. 16

of molecular changes (like receptor or transporter binding density) after therapy. Discovery of the biological foundation for distinct symptoms may be just as or more useful in understanding the pathophysiology of the illness than coercing a collection of symptoms under one biologically likely justification or diagnosis[107].

Going back to the Human Genome project, it is recognized that the vision of P-med stemmed from that research. P-med is the study of delivering the right drug, at the right time, and in the right dose. This is done for an individual patient through the integration of all collected data from the patient, especially genomic information. In short, the purpose of P-med is to answer the question "which intervention is the best one?" by the process of understanding target, characteristic features that lead to a certain treatment, treatment management, and use of the least harmful therapy that creates the best outcome. P-med is about give-and-take for treatment and side-effects of the treatment. As stated before, individualized medicine is most prominent in the oncology field with the help of P-med[108]. Many times, target drugs are found to be useful but cannot get past Phase II of testing because they are considered to be unnecessary and not cost-efficient enough[109]. P-med is based on personalized and systematic clarifications of molecular interactions, metabolic pathways, and biomarkers[110].

[107] Ozomaro, Wahlestedt, & Nemerodd, 2012, p. 15
[108] Nardine, Annoni, & Schiavone, 2012, p. 1001
[109] Nardine, Annoni, & Schiavone, 2012, p. 1002
[110] Nardine, Annoni, & Schiavone, 2012, p. 1004

Another recently developed field of study for psychiatry (and other medical disciplines) involves characterizing endophenotypes. It is a captivating endeavor because this study may present a less multifaceted ailment predecessor that is more feasible for study than the illness syndrome itself. For this process to work, the endophenotypes must be connected to a heritable illness common in a population and be primarily state independent[111]. The availability of any additional, objective measure can help psychiatrists in the diagnostic process which, of course, improves outcome rates and leads to more efficient treatment[112].

The future of personalized medicine in psychiatry essentially echoes ideals that are unrealized. At the moment, the field is at the information gathering stage. The greatest progress will come from bits and pieces of all the systems from above and will inevitably make psychiatry based on biology. But one should not be discouraged at the lack of positive information received thus far, because the newest applications including imaging genomics, show promise in personalizing medicine[113]. Personalized medicine will ensure that the most effective treatment will be used from patient-to-patient and will have a large impact on reducing the cost for medical treatment. Personalized medicine is proclaimed to be the major transformative milestone needed to improve health care[114].

---

[111] Moran, Pol, & Gogtay, 2013, p. 3216
[112] Schnack et al., 2014, p. 299
[113] Ozomaro, Wahlestedt, & Nemerodd, 2013, p. 24
[114] Bombard, Abelson, Simeonov, & Gauvin, 2013, p. 1197

Difficulties

As good as genomics sounds, and despite the endless possibilities of these studies, there are still a few shortcomings to be overcome with the methods. Individualized medicine touches on ethical boundaries for balancing autonomy with resource distribution[115]. When embarking on the journey to personalize medicine, citizen panels had shared expectations for results of improved care[116]. These panels expected increased public awareness and education to increase confidence in the use of the technology. This education would include counseling services to help patients through the process and pointing out many more options available to them for their treatment. At the same time, there is a fear of having too many options, and there is the concern that patients' access to treatment options can become limited. Another worry comes from ethical issues as to when to accept the fact that there are no longer any plausible options left, but the individualized medicine may extend the inevitable, how long should the treatment continue? This issue occurs now, but it is feared that it can become more complicated with more options. In the end, the panels agreed that the treatment decisions of the patient should, at the end of the day, be left to the patient, because the treatment might offer the patient hope[117].

---

[115] Bombard, Abelson, Simeonov, & Gauvin, 2013, p. 1199
[116] Bombard, Abelson, Simeonov, & Gauvin, 2013, p. 1198
[117] Bombard, Abelson, Simeonov, & Gauvin, 2013, p. 1199

Conclusion

Bipolar Disorder and Schizophrenia afflict approximately 2 % of the population worldwide. Because these conditions can be very debilitating to both the victim and those close to them, discerning the cause would benefit an even larger percentage of the population. The most recent research focuses on finding the causes of BPD and Schizophrenia using genetic mapping and brain scans. To date, no underlying cause has yet been found, and no specific gene has been conclusively identified. Lithium remains the best treatment for BDP for now, and it may also be useful for Schizophrenia. Anti-psychotics work best for Schizophrenia. Lithium is an anti-convulsant as well, and most anti-convulsants double as mood stabilizers. Correct diagnoses of BPD versus Schizophrenia or major depression will also result in more effective treatment regiments. At this time doctors can readily identify a psychological impairment, but have difficulty differentiating between the maladies without brain scans and/or genetic mapping. Unfortunately, much more research is necessary to enable these diagnoses. In addition to the genetic information that can be used for diagnoses, environmental factors and comorbidity also play major roles in the timing, the onset, and the severity of both BPD and Schizophrenia. Anxiety and drug use hide the symptoms and make diagnoses even more difficult to pinpoint. The new wave of genetic research is showing great promise in cancer treatment, and will undoubtedly soon also benefit victims of psychological disorders and their families.

References

Alda, M., Grof, P., Rouleau, G. A., Turecki, G., & Young, L. T. (2005). Investigating
responders to Lithium prophylaxis as a strategy for mapping susceptibility
genes for Bipolar Disorder. *Progress in Neuro-psychopharmacology &
Biological Psychiatry, 29,* 1038-1045.

Andreasen, N. (2005). Symptoms, signs, and diagnosis of
Schizophrenia. *Lancet, 346*(8973), 1-12.

Bombard, Y., Abelson, J., Simeonov, D., & Gauvin, F. (2013). Citizens'
perspectives on personalized medicine:A qualitative public deliberation
study. *European Journal of Human Genetics, 21,* 1197-1201.

Clarke, M. C., Harley, M., & Cannon, M. (2005). The Role of Obstetric Events in
Schizophrenia. *Schizophrenia Bulletin, 32*(1), 3-8.

Demjaha, A., MacCabe, J., & Murray, R. (2011). How genes and environmental
factors determine the different neurodevelopmental trajectories of
Schizophrenia and Bipolar Disorder. *Schizophrenia Bulletin, 38*(2), 209-214.

Holsboer, F. (2008). How can we realize the promise of personalized antidepressant
medicines? *Nature Reviews Neuroscience, 9*(8), 638-646

Lachman, H. M. (2000). An overview of Bipolar Disorder: An inherited psychiatric
disorder. *The Einstein Quarterly Journal of Biology and Medicine, 17*(3),
135-142.

Lander, E. S. (2011). Initial impact of the sequencing of the human genome. *Nature*,
1-43

Malhi, G., Bargh, D., Cashman, E., Frye, M., & Gitlin, M. (2012). The clinical
management of Bipolar Disorder complexity using a stratified model. *Bipolar
Disorders, 14*(2), 66-89.

McClellan, J., & Werry, J. (2013). Practice Parameters for the Assessment and
Treatment of Children and Adolescents with Schizophrenia. *Journal of The
American Academy of Child and Adolescent Psychiatry, 52*(9), 976-987.

Moran, M. E., Pol, H. H., & Gogtay, N. (2013). A family affair: Brain abnormalities in
siblings of patients with Schizophrenia. *Brain, 136*, 3215-3226.

Nardini, C., Annoni, M., & Schiavone, G. (2012). Mechanistic understanding in
clinical practice: complementing evidence-based medicine with personalized
medicine. *Journal of Evaluation in Clinical Practice, 18*, 1000-1005.

Nav Haren, N., Rijsdijk, F., Schnack, H., Picchioni, M., & Toulopoulou, T. (2012).
The genetic and environmental determinants of the association between
brain abnormalities and Schizophrenia: The Schizophrenia twins and
relatives consortium. *Society of Biological Psychiatry, 71*, 915-921.

Offord, J. Genetic approaches to a better understanding of Bipolar Disorder.
Pharmacology & Therapeutics, 133, 133-141.

Ozomaro, U., Wahlestedt, C., & Nemeroff, C. (2013). Personalized medicine in psychiatry: problems and promises. *BMC Medicine*, *11*(132), 1-35.

Porteous, D. J., Evans, K. L., Millar, J. K., Pickard, B. S., Thomson, P. A.,

James, R., Blackwood, D. H. (2003). Genetics of Schizophrenia and Bipolar

Affective Disorder: Strategies to identify candidate genes. *Cold Spring*

Harbor Symposia on Quantitative Biology, *68*, 383-391.

Roofeh, D., Tumuluru, D., Shilpakar, S., & Nimgaonkar, V. (2013). Genetics of

Schizophrenia: Where has the heritability gone? *International Journal of*

Mental Health, *42*(1), 5-22.

Schnack, H., Nieuwenhuis, M., & Van Haren, N. (2014). Can structural MRI aid in

clinical classification? A machine learning study in two independent samples

of patients with Schizophrenia, Bipolar Disorder, and healthy subjects.

NeuroImage, *84*, 299-306.

Shaltiel, G., Chen, G., & Manji, H. K. (2007). Neurotrophic signaling cascades in the

pathophysiology and treatment of Bipolar Disorder. *Current Opinion in*

Pharmacology, *7*, 22-26.

Turecki, G., Grof, P., Grof, E., D?Souza, V., Lebuis, L., Marineau, C., . . . Alda, M.

(2001). Mapping susceptibility genes for Bipolar Disorder: A

pharmacogenetic approach based on excellent response to Lithium.

Molecular Psychiatry, 6, 570-578.

Van Haren, N., Cahn, W., Hulshoff Pol, H., & Kahn, R. (2012). The course of brain abnormalities in Schizophrenia: Can we slow the progression? *Journal of Psychopharmacology, 26*(5), 8-14.